Inspire Science

Be a Scientist
Notebook

Student Journal

Grade K

Mc
Graw
Hill
Education

mheducation.com/prek-12

STEM McGraw-Hill is committed to providing
instructional materials in Science, Technology,
Engineering, and Mathematics (STEM) that give all
students a solid foundation, one that prepares them
for college and careers in the 21st century.

Send all inquiries to:
McGraw-Hill Education
STEM Learning Solutions Center
8787 Orion Place
Columbus, OH 43240

ISBN: 978-0-07-678219-2
MHID: 0-07-678219-0

Printed in the United States of America.

9 10 11 LWI 24 23

Our mission is to provide educational resources that enable
students to become the problem solvers of the 21st century
and inspire them to explore careers within Science, Technology,
Engineering, and Mathematics (STEM) related fields.

TABLE OF CONTENTS

KAYLA
Landscape Architect

TABLE OF CONTENTS

Check out the activities in every lesson!

KAYLA
Landscape Architect

Inspire Science

This is your journal. You will be a scientist or an engineer. Use your journal to help you answer questions and solve problems.

Draw a picture to show what a scientist or an engineer might do.

ERIC
Video Game Designer

Forces and Motion

How can trains change speed and direction?

FINN
Construction Manager

 ## Science in My World

Look at the train. What do you wonder about how the train moves? Draw a picture. Show how you would move the train.

Science and Engineering Practices

I will plan an investigation.
I will carry out an investigation.

Pushes and Pulls

 PAGE KEELEY SCIENCE PROBES

Push or Pull?

What is Paige doing?

☐ Paige is pushing the wagon.

☐ Paige is pulling the wagon.

💬 Talk about your thinking with a partner.

🌍 Science in My World

▶ Watch the video. You see a hammer. Think about what the hammer did.

What do you wonder about how a hammer works?

> I wonder how hammers make things move.

CHLOE
Carpenter

❓ Essential Question

How do pushes and pulls affect the way objects move?

Draw a push.	Draw a pull.

⚙️ Science and Engineering Practices

I will plan an investigation.
I will carry out an investigation.

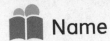

Name _____

 Inquiry Activity
Push and Pull

Make a Prediction (Circle) the answer.

Which will be easier to move?

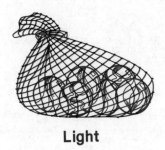

Light

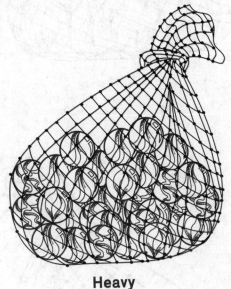

Heavy

Carry Out an Investigation

Move the bags. See what is different.

Push and pull each bag.

Record Data

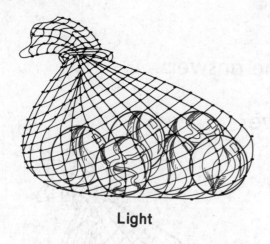

Light

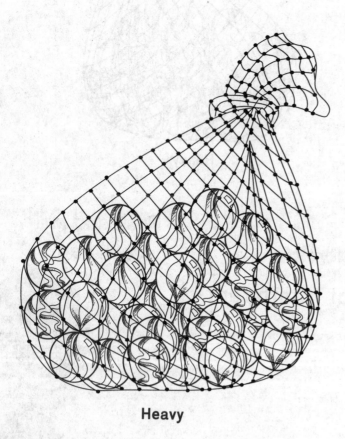

Heavy

Which bag was easier to push?

Circle the bag that was easier to push.

How did your pushes and pulls change? Tell the class.

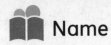 **Name** _____

🔤 Vocabulary

Listen for these words as you learn
about pushes and pulls.

motion	distance	force
position	pull	push
speed		

Pushes and Pulls

📖 Read.

🗨 Talk about it. (Circle) two words.
Draw a picture of the words.

Push, Pull, Collide

▶ Watch the video.
Circle the word that matches each picture.

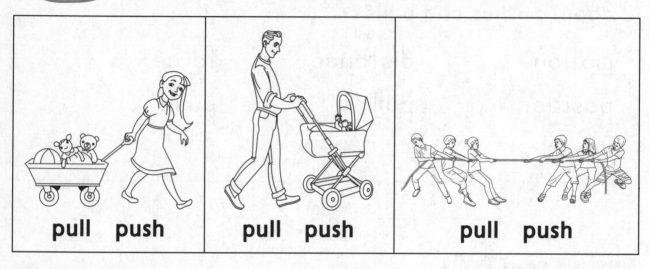

pull   push   pull   push   pull   push

Pushes and Pulls

🔄 Explore to learn about pushes and pulls.
Circle the push pictures.

Pushing and Pulling

👁 Look at the pictures. (Circle) the big push. Put an ✕ on the small push.

⚙ Crosscutting Concepts
Cause and Effect

Look at the picture. Do you see a push or a pull? (Circle) push or pull.

push   pull

⚙ Science and Engineering Practices

Say the **I can** statements together.
Tell a partner what you can do.

I can plan an investigation.
I can carry out an investigation.

✋ Inquiry Activity
Push and Pull

Make a Prediction (Circle) the answer.

1. What will happen when you pull
   a wagon?

   It will come toward you.

   It will go away from you.

2. What will happen when you push
   a wagon?

   It will come toward you.

   It will go away from you.

Carry Out an Investigation

▦ Play. Make the monkey push
and pull a wagon.

Help the monkey
get the dog to
the finish line.

Record Data Record what happens.
Draw pictures.

I made the monkey **pull** the wagon.

I made the monkey **push** the wagon.

💬 What moved the wagon more?
Tell a partner.

⚙️ Performance Task
A Moving Car

Materials
☐ toy car
☐ string or yarn

Directions

① Use a push to make the car move.

② Use a pull to make the car move.

③ Draw a push.

④ Draw a pull.

Record Data

Push	Pull

💬 What made the car move?
Tell a partner.

? Essential Question

How do pushes and pulls affect the way objects move?

💬 Talk about the question as a class. Write the answer.

_____

- -

_____

- -

_____

- -

_____

⚙ Science and Engineering Practices

I **did** plan an investigation.
I **did** carry out an investigation.

Now that you're done with the lesson, rate how well you did.

Rate Yourself

Name _____

When Objects Collide

 PAGE KEELEY
SCIENCE
PROBES

Toy Car Crash

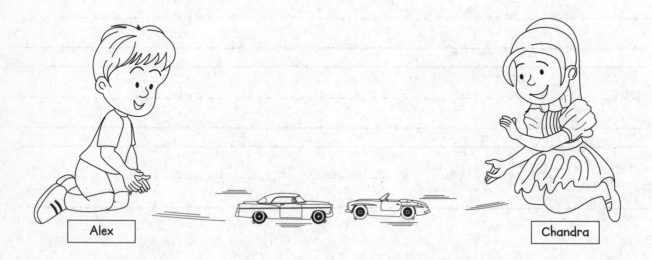

Alex

Chandra

What will happen when the toy cars crash into each other? Who has the best idea?

Alex: *They will move in the same direction.*

Chandra: *They will move in a different direction.*

💬 Talk about your thinking with a partner.

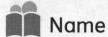
Science in My World

▶ Watch the video. See the ball move. Think about how it moves. What do you wonder?

I wonder how things move when something else hits them.

RILEY
Automotive Engineer

? Essential Question
What happens when objects touch or collide?

Circle pictures that show objects that have been touched or hit.

Science and Engineering Practices

I will carry out an investigation.

✋ Inquiry Activity
Marbles Collide

Make a Prediction What will happen when two marbles bump together? Draw your prediction.

Carry Out an Investigation

Push one marble into another. Try again with a different marble.

Bump the marbles together. See what happens.

Record Data What happened when two marbles bumped together? Draw what you saw.

Roll	2 small marbles	1 big marble 1 small marble
Roll 1		
Roll 2		

Name _____

abc Vocabulary

> Listen for this word as you learn about when things collide.
>
> collide

When Things Collide

📖 Read.

💬 Talk to a partner. Use the new word.
Draw a picture.

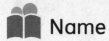

When Objects Collide

▶ Watch the video. What will happen?

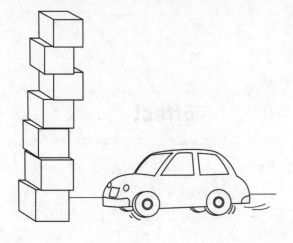

1. The blocks will ___ fall ___.

Objects Colliding

🔊 Explore. Draw a picture.

⚙ Crosscutting Concepts
Cause and Effect

Draw what will happen when objects collide.

Cause	Effect

⚙ Science and Engineering Practices

Say the **I can** statement together.
Tell a partner what you can do.

I can carry out an investigation.

 Inquiry Activity
Bottle Bowling

Make a Prediction What will happen when you give the ball a small push?

(Circle) the answer.

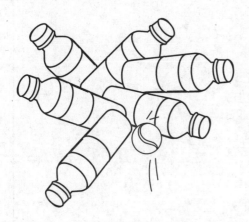

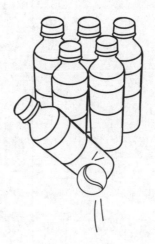

Carry Out an Investigation

Push the ball into the bottles. Try again with a harder push.

Push the ball different ways. What happens?

Record Data Draw what you saw.

Small Push	Big Push

💬 Sometimes the ball knocked down many bottles. Sometimes it knocked down a few. Why? Tell a partner.

 Performance Task
Balls Colliding

Materials

- ☐ basketball
- ☐ playground ball
- ☐ tennis ball
- ☐ ping pong ball

Make a Prediction

What will happen when two different balls collide?

Directions

1. Choose two balls.

2. Push them into each other.

3. See what happens to the balls.

4. Try it again with different balls.

5. Draw what you see.

Watch what happens when two different balls collide.

Record Data Draw what happens to the balls.

First 2 balls	Next 2 balls

💬 Talk to a partner about how the different balls moved.

? Essential Question
What happens when objects touch or collide?

💬 Talk about the question as a class.
Write the answer.

_____

- -

_____

- -

_____

- -

_____

⚙ Science and Engineering Practices

I did carry out an investigation.

Rate Yourself

Now that you're done with the lesson, rate how well you did.

Name _____

Direction and Force

PAGE KEELEY
SCIENCE
PROBES

Changing Direction

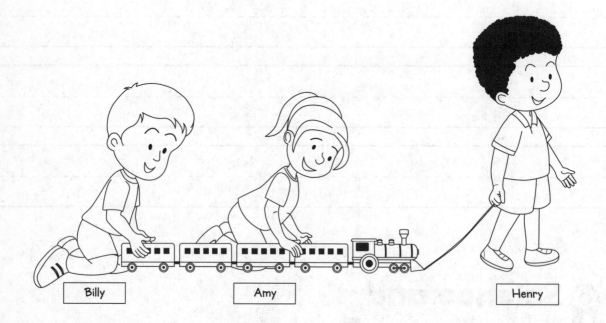

| Billy | Amy | Henry |

How can you change the direction of the train? Who do you agree with?

Billy: *Give it a push.*

Henry: *Give it a pull.*

Amy: *Give it a push or a pull.*

Talk about your thinking with a partner.

 Science in My World

▷ Watch the video. Think about
how the baseball moves.
What do you wonder about
how the bat moves the ball?

*I want to
learn how to change
the way an object moves.*

CJ
Statistician

? Essential Question
How can pushes and pulls change an object's direction?

Push	Pull

Science and Engineering Practices

I will analyze data.

🖐 Inquiry Activity
Changing the Way an Object Goes

Make a Prediction Draw an arrow to show where you think the ball will go.

> Try changing the direction of the ball.

Carry Out an Investigation

Change the direction the ball moves. Try a push or a pull.

Name _____

Record Data Draw how you moved the ball. Use arrows to show how the direction changed.

abc Vocabulary

Listen for this word as you learn about directions and force.

direction

Change Directions

Explore.

Why do objects change directions? Draw a picture of how an object changes direction. Use an arrow to show the direction the object moves.

Marble Maze

▶ Watch the video.
How did the marble in the maze move?
Make a drawing.

Pushes and Pulls

📖 Read.

💬 Talk about it. Use the word *direction*.

⚙ Science and Engineering Practices

Say the **I can** statement together.
Tell a partner what you can do.

I can analyze data.

✋ Inquiry Activity
Make a Marble Maze

How will the marble move?
Draw arrows in the maze.

Can you make your marble move in different directions?

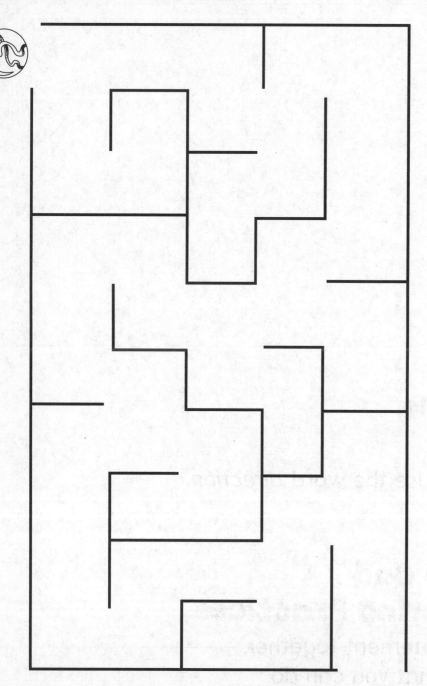

 **Name** _____

Carry Out an Investigation

1. Build a maze.

2. Move a marble through the maze.

Record Data Draw your maze.
Draw arrows to show the path
the marble took.

Name _____

⚙ Performance Task
Using a Pulley

Materials

☐ rope

☐ small bag or basket with handles

☐ two chairs

Directions

① Make a pulley system.

② Move the basket in one direction.

③ Draw what you saw.

④ Make the basket move in another direction.

⑤ Draw what you saw.

Record Data Draw how you made the basket change direction.

? Essential Question
How can pushes and pulls change an object's direction?

💬 Talk about the question as a class.
Write the answer.

_____

- -

_____

- -

_____

- -

_____

⚙️ Science and Engineering Practices

I did analyze data.

Rate Yourself

Now that you're done with the lesson, rate how well you did.

Name _____

Forces and Motion

⚙ Performance Project
Design a Solution

What ways can you make the train go faster? How can you change the train's direction? Design a way for Finn to quickly move the train over and around objects.

Plan an Investigation

Draw the materials you will need.

[blank box]

💬 How will your solution change the speed and direction of a train? Tell a partner.

A train changes speed when it is going down a hill.

Draw a picture of your solution.

Name _____

Read Together

Energy and the Sun

How can we keep the playground from getting too hot?

CHLOE
Carpenter

 ## Science in My World

It is a sunny day. You see the playground. What does the Sun shining on it make you wonder?

Draw a picture of your thought.

Science and Engineering Practices

I will carry out an investigation.
I will design a solution.
I will make a model.

Sunlight and Earth's Surface

 **PAGE KEELEY SCIENCE PROBES**

Warm Sand

Linda   Wei   Olivia

Why does the sand feel warm?
Who do you agree with?

Linda: *The wind warms the sand.*

Wei: *The sun warms the sand.*

Olivia: *The water warms the sand.*

💬 Talk about your thinking with a partner.

Science in My World

▶ Watch the video. What do you see?
Draw a picture.

[blank drawing box]

? Essential Question
How does the Sun affect Earth's surface?

What happens on Earth when the Sun rises?

HUGO
Meteorologist

⚙ Science and Engineering Practices
I will carry out an investigation.

✋ Inquiry Activity
Sunlight and Water

Make a Prediction Will sunlight make
water warmer? (Circle) your answer.

Yes   No

	Water	Water in Sunlight
Start		
End		

💬 **What happened to the water?**
Talk about it with a partner.

⚙ Crosscutting Concepts
Cause and Effect

Circle the thermometer that has been in the Sun.

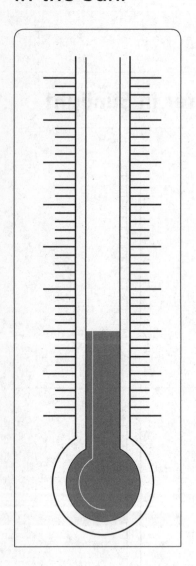

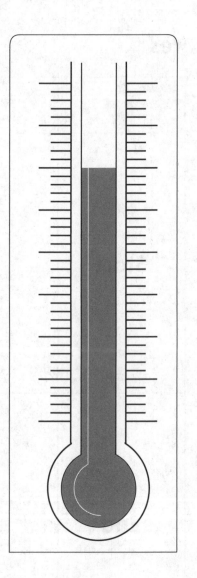

 Name _____

abc Vocabulary

Listen for these words as you learn about sunlight and Earth's surface.

Earth   heat   Sun

temperature   warm

The Sun Throughout the Day

▶ Watch the video.

💬 Talk to a partner. (Circle) two words.
Use the new words. Draw a picture.

Earth and the Sun

📖 Read.

💬 Talk to a partner. Draw how Earth goes around the Sun.

💬 What does the Sun do for Earth?

⚙️ Science and Engineering Practices

Say the **I can** statement together.
Tell a partner what you can do.

I can carry out an investigation.

🖐 Inquiry Activity
Sunlight and Earth's Surface

Predict Which things will the Sun warm?
Circle them.

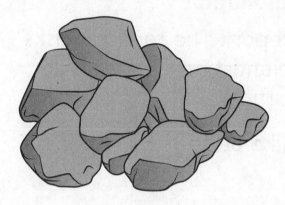

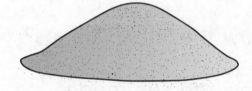

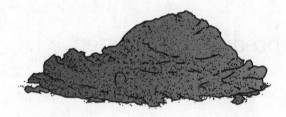

Carry Out an Investigation

Touch the things.

Leave them in the Sun.

Touch them again.

💬 What happened? Did you predict this?

Name _____

Performance Task
Give a News Report

Give a news report about the Sun.

Construct an Explanation

Act out a news report. The reporter asks questions. The scientist answers them. Use what you learned.

1. What was it like before the Sun came up?

2. What happened next?

3. How do you know the Sun lit things?

4. How do you know the Sun made things warm?

5. What do you predict will happen tonight?

Crosscutting Concepts
Cause and Effect

💬 What is different when the Sun is shining and when it is not?

? Essential Question
How does the Sun affect Earth's surface?

💬 Talk about the question as a class. Write the answer.

_____
- -
_____
- -
_____
- -
_____

⚙ Science and Engineering Practices

I did carry out an investigation.

Rate Yourself

Now that you're done with the lesson, rate how well you did.

Sunlight and Shade

Sunlight and Shade

Kenny   Jim

Kenny and Jim are playing outside.
It is hot and sunny.

Which boy has the best idea?

Kenny: *Let's play in the shade to stay cool.*

Jim: *The shade is darker, not cooler.*

💬 Talk about your thinking with a partner.

Name _____

 ## Science in My World

Look at the photo. What do you see? What do you think it feels like under the umbrellas? What would it feel like without umbrellas? Draw pictures.

? Essential Question
How can we stay cool in the Sun?

What does the umbrella do to sunlight?

HANNAH
Welder

⚙ Science and Engineering Practices

I will design a solution.
I will make a model.

✋ Inquiry Activity
Temperatures Throughout the Day

Carry Out an Investigation

▦ Explore to learn about shade.

💬 How did the temperature change? Why?

How did the temperature change? What caused the change?

⚙ Crosscutting Concepts
Cause and Effect

Which makes it coolest?
Circle the picture.

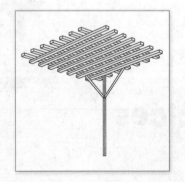

abc Vocabulary

Listen for these words as you learn about sunlight and shade.

cool   shade

Earth and the Sun

📖 Read.

💬 Talk to a partner. Use the new words.

Shade from the Sun

▶ Watch the video. Draw what you see.

How are things different in sunlight and in shade? Draw your thoughts.

Sun	Shade

Science and Engineering Practices

I can design a solution.

I can make a model.

 Name _____

Inquiry Activity
Animal Shelters

Make a Model

1. Find a picture of an animal in a magazine.

2. Cut out the picture of the animal.

3. Glue it on paper.

4. Draw a shelter from the Sun for the animal.

5. Make a class book.

💬 Show your page to the class. How does your shelter help your animal? Explain.

Crosscutting Concepts
Cause and Effect

💬 How does your shelter cool an animal? Talk to a partner.

Name _____

⚙ Performance Task
Draw an Animal Shelter

Choose an animal. Circle your choice.

Draw a shelter for your animal.

 **Name** _____

? Essential Question
How can we stay cool in the Sun?

💬 Talk about the question as a class.
Write the answer.

_____

- -

_____

- -

_____

- -

_____

⚙️ Science and Engineering Practices

I did design a solution.
I did make a model.

Rate Yourself

 Now that you're done with the lesson, rate how well you did.

Name _____

Energy and the Sun

⚙ Performance Project
Design a Structure to Make Shade

It is a sunny day. You are on the playground. Look for a place that needs shade. How can you build something to make shade?

Draw what you will need in the box.

How can you make
the playground cooler?

What did you build? Draw a picture.

💬 **Did you make shade? Talk about it.**

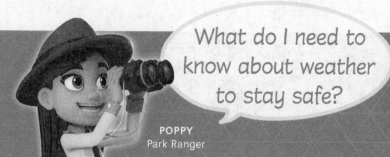

Weather

> What do I need to know about weather to stay safe?

POPPY
Park Ranger

 ## Science in My World

▶ Watch the video. You see lightning. Think about when lightning happens. What do you wonder about lightning?

💬 How can Poppy tell what the weather will be like tomorrow? Explain your thinking to your class.

 ## Science and Engineering Practices

I will communicate information.
I will develop a model.
I will analyze data.

Describe Weather

 PAGE KEELEY SCIENCE PROBES

Thermometer

Sonia   Ava   Wendy

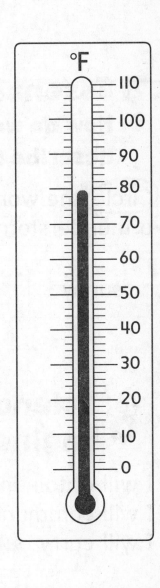
°F
110
100
90
80
70
60
50
40
30
20
10
0

A thermometer is a weather tool.
What does a thermometer measure?
Who has the best idea?

Sonia: *It measures the rain.*

Ava: *It measures the temperature.*

Wendy: *It measures how strong the
       wind blows.*

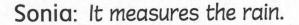 **Talk about your thinking with a partner.**

Science in My World

 You see rain in a jar. Think about how it got in the jar. What do you wonder about rain?

> I wonder how tools can help describe weather.

? Essential Question

How do we measure and describe weather?

Circle the word that describes a thunderstorm.

sunny	rainy	snowy	dry

HUGO
Meteorologist

⚙ Science and Engineering Practices

I will obtain information.

I will communicate information.

I will carry out an investgation.

✋ Inquiry Activity
Rain Gauge Simulation

Make a Prediction (Circle) the answer.

What will fill the rain gauge the most?

Change the time and the weather. Then watch the rain gauge.

Carry Out a Simulation

▦ Explore to learn about a rain gauge.

Record Data Circle the answer.

1. Which filled the rain gauge the most?

2. What happened when it rained for more hours?

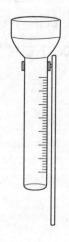

🗨 What patterns did you notice?
Tell your class.

Inquiry Activity
Weather Graph

Make a Prediction Think about the next three days. What do you think the weather will be? Draw pictures.

Day 1	
Day 2	
Day 3	

Carry Out an Investigation

Record the weather for three days.

Look outside to see the weather.

Record Data Make an ✕ to show the weather.

	Day 1	Day 2	Day 3
Sunny			
Windy			
Rainy			
Snowy			

🔤 Vocabulary

Listen for these words as you learn
how to describe weather.

temperature   thermometer   cool

rain   weather   warm

rainbow   clouds

Weather and Seasons

📖 Read.

💬 Talk about it. (Circle) two words.
Draw a picture of the words.

✋ Inquiry Activity
Measure the Temperature

Use a tool to measure the temperature.

Make a Prediction (Circle) the place you think will be warmer.

inside   outside

Carry Out an Investigation

Which tool measures temperature? (Circle) it.

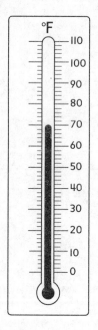

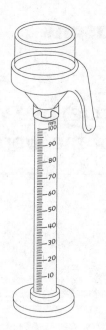

Record Data Which thermometer shows the inside temperature? Which shows the outside temperature? (Circle) *inside* or *outside.*

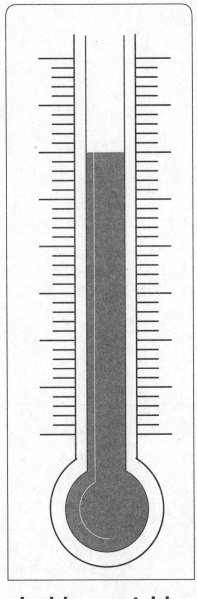

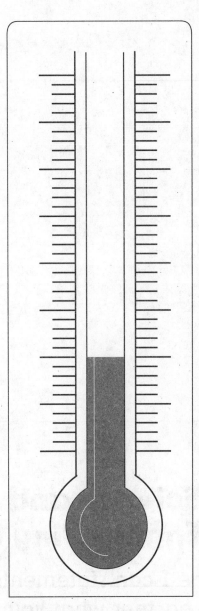

inside   outside   inside   outside

💬 How are the temperatures inside and outside different? Tell the class.

Name _____

Kinds of Weather

▶ Watch the video. What is the weather today? Draw a picture.

⚙ Science and Engineering Practices

Say the **I can** statements together.
Tell a partner what you can do.

I can obtain information.
I can communicate information.
I can carry out an investigation.

 Name _____

ELABORATE

Inquiry Activity
Wind Effects

Make a Prediction

1. Choose two objects.

2. Draw them in the first box.

3. Draw what you think you will see in the second box.

Object	I Predict	What I Saw

Carry Out an Investigation

What happened to each object?

4. Draw what you saw in the third box.

💬 What happens to objects when it is windy? Tell the class.

Online Content at connectED.mcgraw-hill.com

Lesson 1 Describe Weather **69**

Name _____

⚙ Performance Task
Make a Weather Poster

① Make a poster about weather and weather tools.

② Copy the words on your poster.

③ Cut out pictures of weather.

④ Glue the pictures on your poster.

⑤ Share it with the class.

Materials

☐ poster board

☐ magazines

☐ scissors

☐ glue sticks

☐ markers

Make a Model

Thermometer	Rain Gauge

? Essential Question

How do we measure and describe weather?

💬 Talk about the question as a class.
Write the answer.

_____

- -

_____

_____

- -

_____

_____

- -

_____

⚙️ Science and Engineering Practices

I did obtain information.
I did communicate information.
I did carry out an investigation.

Now that you're done with the lesson, rate how well you did.

Rate Yourself

Weather Patterns

PAGE KEELEY
SCIENCE
PROBES

Weather Patterns

| Maria | Omar | Erin |

Three friends are talking about weather. Who is describing a weather pattern?

Omar: It is very cold today.

Maria: January is colder than June.

Erin: It snowed for five hours.

💬 Talk about your thinking with a partner.

Name _____

 Science in My World

▶ Watch the video. You see weather in the seasons. Think about the seasons. What do you wonder about weather in the seasons?

I wonder if there are patterns in the weather.

HUGO
Meteorologist

? Essential Question
What weather patterns do you observe in the seasons?

Which season do you see?
Trace the word.

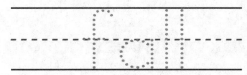

⚙ Science and Engineering Practices

I will interpret data.
I will develop a model.

Inquiry Activity
Nature Walk

Ask a Question How can you tell the season? Draw three clues.

Clue 1	Clue 2	Clue 3

Carry Out an Investigation

You are the detective! What clues can you find?

What clues tell you about the season?

Record Data Draw pictures to show the clues you found.

abc Vocabulary

Listen for these words as you learn about weather patterns.

| patterns | season | spring |
| summer | fall | winter |

Weather and Seasons

📖 Read.

As you listen, think about patterns in the seasons.

⚙ Crosscutting Concepts
Patterns

💬 What pattern do the seasons make? Tell a partner. Use the new words.

Seasons

🔊 Explore to learn about spring and summer. Copy the words.

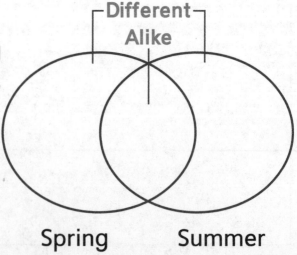

┌Different┐
Alike

Spring   Summer

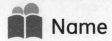

✋ Inquiry Activity
Five-Day Forecast

Interpret Data Look at the forecasts. How are they alike? How are they different? Draw a picture.

How are the forecasts alike and different?

⚙ Crosscutting Concepts
Patterns

(Circle) the season shown in each picture.

spring summer fall winter	spring summer fall winter
spring summer fall winter	spring summer fall winter

⚙ Science and Engineering Practices

Say the **I can** statements together.
Tell a partner what you can do.

I can interpret data.
I can develop a model.

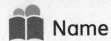 Name _____

Seasons in Different Places

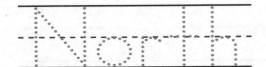 Explore.

Think about seasons in other states. Where would you like to live? Trace your answer.

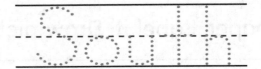 North South

 Inquiry Activity
Favorite Season

Interpret Data Ask: What is your favorite season? Use an ✕ for each answer.

Fall	
Winter	
Spring	
Summer	

⚙ Performance Task
Make a Seasons Foldable

Materials

☐ paper

☐ crayons or colored pencils

Use what you know about the different seasons to make a model. Draw pictures.

Make a Model

Fold the paper. Label it. Draw pictures.

spring	summer
fall	winter

💬 What is the same in each season?
What is different? Tell a partner.

❓ Essential Question
What weather patterns do you observe in the seasons?

💬 Talk about the question as a class.
Write the answer.

_____

- -

_____

_____

- -

_____

_____

- -

_____

⚙️ Science and Engineering Practices

I did interpret data.
I did develop a model.

Now that you're done with the lesson, rate how well you did.

Rate Yourself

Name _____

Forecasting and Severe Weather

PAGE KEELEY
SCIENCE
PROBES

Forecast

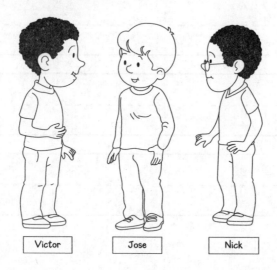

Victor   Jose   Nick

A weatherperson forecasts the weather.
What does it mean to forecast the weather?
Who has the best idea?

Victor: *Describe what the weather will be like next year.*

Jose: *Describe what the weather will be like tomorrow.*

Nick: *Describe what the weather is like right now.*

💬 Talk about your thinking with a partner.

 Name _____

Science in My World

▶ Watch the video. Think about how you can prepare and stay safe in a thunderstorm. What do you wonder about preparing for thunderstorms?

? Essential Question
What does a weather forecast tell us about severe weather?

💬 Talk about your answer with the class.

I wonder how people predict severe weather.

HUGO
Meteorologist

⚙️ Science and Engineering Practices

I will analyze data.
I will communicate information.

Inquiry Activity
Tomorrow's Weather

Make a Prediction What will the weather be like tomorrow? Draw a picture.

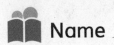
Carry Out an Investigation

Record the weather during the day.

Look outside to see the weather.

Record Data Draw the weather at each time of day.

Morning	Noon
Afternoon	**Evening**

abc Vocabulary

Listen for these words as you learn
about forecasting and severe weather.

forecast   severe weather   hurricane

blizzard   thunderstorm   tornado

Storm Warning

📖 Read.

💬 Talk about severe weather with your
class. (Circle) two words. Use the
new words.

Tornado Warning

▶ Watch the video. Meteorologists
predict tornadoes. What tool do they
use? (Circle) the answer.

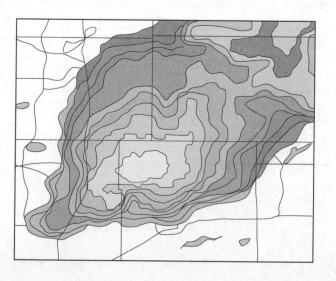

Meteorologists warn people about tornadoes. What tools do they use? Circle the answers.

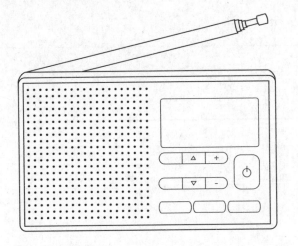

What should you do to prepare for severe weather?

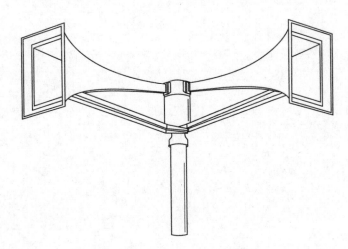

Name _____

Kinds of Severe Weather

▶ Watch the video. What kinds of severe weather happen where you live? Draw a picture.

💬 How can you tell if severe weather is coming?

Severe Weather

📖 Read.

How can you prepare for a rainy day?
Draw your ideas.

How can you prepare for a snowy day?
Draw your ideas.

Forecasting Weather

🔎 Explore.

How do meteorologists help people?
Draw a picture.

⚙ Science and Engineering Practices

Say the **I can** statements together.
Tell a partner what you can do.

I can analyze data.
I can communicate information.

Types of Severe Weather

🔊 Explore.

What type of severe weather is shown in each picture? (Circle) the answer.

blizzard   **hurricane**

tornado   **thunderstorm**

blizzard   **hurricane**

tornado   **thunderstorm**

blizzard   **hurricane**

tornado   **thunderstorm**

blizzard   **hurricane**

tornado   **thunderstorm**

Tools for Severe Weather

🔁 Explore.

(Circle) the items that are useful in severe weather.

flashlight	batteries	soccer ball
canned food	nail polish	water bottle
rubber duck	radio	guitar

💬 Talk about these tools with a partner.

Impact of Severe Weather

🔊 Explore.

What are some problems caused by severe weather?

⚙ Crosscutting Concepts
Cause and Effect

Match the pictures.

rain

snow

wind

⚙ Performance Task
Make a Video

Use what you know about forecasting severe weather to make a video.

Communicate Information

1 Choose a blizzard, hurricane, thunderstorm, or tornado.

2 Make a forecast for that weather.

3 Play a weather forecaster on TV.

4 Tell people about the forecast.

 Name _____

? Essential Question
What does a weather forecast tell us about severe weather?

🗨 Talk about the question as a class.
Write the answer.

_____

- -

_____

- -

_____

- -

_____

 # Science and Engineering Practices

I did analyze data.
I did communicate information.

> Now that you're done with the lesson, rate how well you did.

Rate Yourself

Copyright © McGraw-Hill Education

Online Content at 🔗 connectED.mcgraw-hill.com

Lesson 3 Forecasting and Severe Weather **95**

Weather

⚙️ Performance Project
Create a Weather Poster

Use what you have learned about weather to make a poster.

Poppy will use the poster to prepare for the birthday party.

Make a Model

1. Draw pictures or cut out pictures from magazines of weather.

2. Glue them on your poster.

3. Show your poster to the class.

Materials

☐ poster board

☐ magazines

☐ scissors

☐ glue sticks

☐ markers

I paid attention to weather patterns and the forecast to stay safe.

Draw Poppy preparing for the party.

💬 **How did knowing about weather help Poppy? Tell a partner.**

Plants and Animals

How do animals and plants live and grow together?

RUBY
Veterinarian

Science in My World

▶ Watch the video. You see a camel.
Think about where the camel lives.
What do you wonder about the camel?

💬 Where does the camel get what it
needs? Tell a partner.

 Science and Engineering Practices

I **will** analyze data.
I **will** interpret data.
I **will** develop models.
I **will** use models.

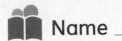 Name _____

Plant and Animal Needs

 **PAGE KEELEY SCIENCE PROBES**

Plant and Animal Needs

(Circle) the things that both plants and animals need to live.

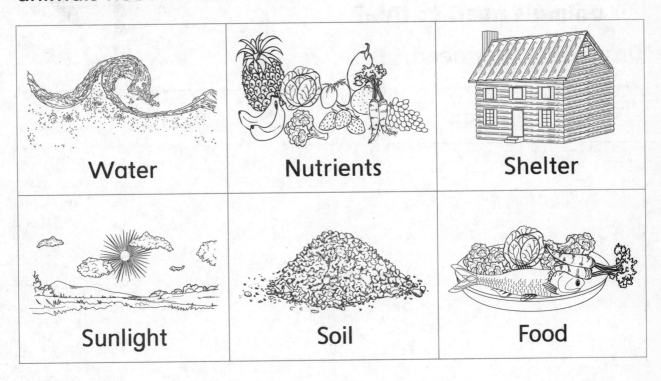

Water

Nutrients

Shelter

Sunlight

Soil

Food

💬 Talk about your thinking with a partner.

Online Content at ➤ connectED.mcgraw-hill.com

Science in My World

▶ Watch the video.
You see a chipmunk. What did it do?
What does it make you wonder?

I wonder how you take care of an animal.

POPPY
Park Ranger

? Essential Question
What do plants and animals need to live?

Draw what they need.

Animals	Plants

⚙ Science and Engineering Practices

I will analyze data.
I will interpret data.

Inquiry Activity
Plant and Animal Needs

Make a Prediction (Circle) the answer.

What does a plant need to grow?

water, sunlight, air

soil, shelter, air

Carry Out an Investigation

Put each plant in a different place. Give each plant water.

Do plants need sunlight to live and grow?

Record Data Draw the two plants.

Sunlight	No Sunlight

Math Connection

Interpret Data Measure how tall the plants are. Use cubes. Draw the cubes next to the plants.

💬 What happened to each plant? Why? Tell a partner.

🔤 Vocabulary

> Listen for these words as you learn
> about plant and animal needs.
>
> | | | |
> |---|---|---|
> | animal | living | need |
> | nonliving | nutrient | plant |
> | shelter | soil | survive |

Plant and Animal Needs

📖 Read.

💬 Talk about it. (Circle) two words.
Use the words.

Rabbit Health

🖩 Explore to learn more about rabbits.

(Circle) things that kept the rabbit healthy.

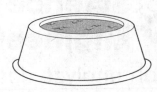

Inquiry Activity
Picture Cards

Make a Prediction Circle your answers.

1. What does a plant need to survive?

   shelter   water

2. What does an animal need to survive?

   food   soil

Carry Out an Investigation

Your teacher will give you a card. What plant or animal is it?

Find a classmate with a card that will help your plant or animal.

What does your plant or animal need to survive?

Science and Engineering Practices

I can analyze data.
I can interpret data.

✋ Inquiry Activity
What Animals Eat

🔊 Explore to learn what animals eat.
Sort the animals.

Make a Model

Draw one of the animals you saw.

What kind of animal did you draw?
Circle the answer.

1. plant eater

2. meat eater

3. eats plants and meats

⚙ Performance Task
Create a Survival Graph

Use what you know about plants and animals to make a survival graph.

Make a Model

① Choose a plant or animal you would like to care for.

② Think about what things your plant or animal needs to survive.

③ Draw your plant or animal on one half of the poster.

④ Draw all the things your plant or animal needs on the same half.

⑤ Draw all the things you need on the other half of the poster.

Plant/Animal	Me

💬 How are the needs of plants and animals like your needs? How are they different?

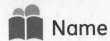
? Essential Question
What do plants and animals need to live?

💬 Talk about the question as a class.
Write your answer.

_____

- -

_____

_____

- -

_____

_____

- -

_____

⚙️ Science and Engineering Practices

I did analyze data.
I did interpret data.

Now that you're done with the lesson, rate how well you did.

Rate Yourself

Name _____

Places Plants Grow

 PAGE KEELEY
SCIENCE
PROBES
Places Plants Grow

Carly

Jaden

Two friends are talking about places where plants grow. Who has the best idea?

Carly: *Plants grow on land.*

Jaden: *Plants grow on land and in water.*

Talk about your thinking with a partner.

🌍 Science in My World

Look at the picture. What do you see? What do you wonder about where plants grow?

I wonder if plants grow on land and in water.

KAYLA
Landscape Architect

❓ Essential Question
Where do different kinds of plants grow?

Draw your ideas.

⚙️ Science and Engineering Practices

I will develop a model.
I will use a model.

🖐 Inquiry Activity
Where Do Plants Grow?

Make a Prediction (Circle) your answers.

1. What kind of plants will you find inside?

   land plants

   water plants

   land and water plants

2. What kind of plants will you find outside?

   land plants

   water plants

   land and water plants

Carry Out an Investigation

Where do you see two different plants growing? What types of plants are they?

Go outside to find out where plants can grow.

Record Data

Where did you see plants grow?
Draw pictures.

Places Plants Grow

Where do plants grow? Why do
they grow there? Tell a partner.

abc Vocabulary

Listen for these words as you learn
about places plants grow.

Arctic   climate   desert
forest   pond

Where Do Plants Grow?

▶ Watch the video.

💬 Talk about the video with a partner.
(Circle) two words. Use the new words.

Draw a picture.

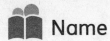 Name _____

✋ Inquiry Activity
Desert or Rainforest Plants

Make a Prediction Circle your predictions.

1. Which plant will survive in the **desert**?

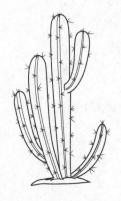 

2. Which plant will survive in the **rainforest**?

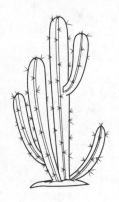 

Learn more about desert and rainforest plants.

Carry Out an Investigation

Find out which plants live in the desert and which plants live in the rainforest.

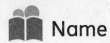

Name _____

Record Data Draw a plant that lives in the desert. Draw one that lives in the rainforest.

Match the Plant to the Climate
Which plant grows in the desert?

The cactus grows in the 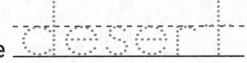_____.

Science and Engineering Practices
I can develop a model.
I can use a model.

✋ Inquiry Activity
Perky Plants

Make a Prediction (Circle) your answer.

Where will the bean plants grow best?

in the sun

in the dark

Carry Out an Investigation

Put the planted beans in different places.
Watch the plants for five days.

Plant one bean
seed in each cup.

Record Data

Draw what you see after five days.

Plant 1	Plant 2	Plant 3

Where did the bean seeds grow best?
Tell a partner.

Performance Task
Perfect Plant

Create a plant that will survive in your climate.

Make a Model

1 Draw the plant.

2 Label the plant parts.

Crosscutting Concepts
Systems and System Models

Why do some plants grow well in some places and not others? Tell a partner.

❓ Essential Question
Where do different kinds of plants grow?

💬 Talk about the question as a class.
Write your answer.

_____
- -
_____
_____
- -
_____

⚙️ Science and Engineering Practices

I did develop a model.
I did use a model.

Now that you're done with the lesson, rate how well you did.

Rate Yourself

Places Animals Live

PAGE KEELEY SCIENCE PROBES

Places Where Animals Live

Three friends are talking about places where animals live. Who has the best idea?

Elle: Animals live on land.

Pete: Animals live in water.

Leroy: Animals live on land and in water.

💬 Talk about your thinking with a partner.

Name _____

🌍 Science in My World

▶ Watch the video. What did you see the deer do? What does it make you wonder?

> I wonder how animals survive in different habitats.

JORDAN
Animal Trainer

❓ Essential Question
Where do different kinds of animals live?

💬 Where do you think deer live? Talk with a partner.

Deer live in the ___forest___.

⚙️ Science and Engineering Practices

I will develop a model.
I will use a model.

Name _____

✋ Inquiry Activity
Animal Homes

Make a Model Choose an animal. How can you build a home for the animal? Draw a picture.

Where does your animal live?

Record Data Draw your animal home.

abc Vocabulary

Listen for these words as you learn
about places animals live.

ecosystem   habitat

Animal and Plant Habitats

📖 Read.

💬 Talk about it. Use the new words.

Where Do Animals Live?

▶ Watch the video.

Draw a line from the animal to the habitat.

arctic

grassland

forest

Inquiry Activity
Rainforest and Desert Animals

Ask a Question Circle your answers.

1. Which animal lives in the rainforest?

2. Which animal lives in the desert?

Carry Out an Investigation

Play a game with the picture cards.

Record Data Draw an animal. Show how the animal can survive.

Desert Animal	Rainforest Animal

⚙ Science and Engineering Practices

I can develop a model.

I can use a model.

Inquiry Activity
Things Humans Need

Think about where you live. What things do you need to survive?

Make a Model

Work with a partner. Draw the things that humans need to survive.

Choose two things that humans need. How do humans get those things? Talk about it with your partner.

⚙ Performance Task
Habitat Model

Use what you know about habitats to make a model.

Make a Model

① Choose a habitat.

② Draw a model of the habitat.

③ Cut out pictures of animals for your habitat.

④ Use classroom materials to build your habitat.

⑤ Place animals in habitat.

⑥ Share with your class.

? Essential Question
Where do different kinds of animals live?

🗨 Talk about the question as a class. Write your answer.

_____

- -

_____

_____

- -

_____

- -

_____

⚙ Science and Engineering Practices

I **did** develop a model.
I **did** use a model.

Rate Yourself

Now that you're done with the lesson, rate how well you did.

Name _____

Plants and Animals

⚙️ Performance Project
Make a Diorama

What materials will you need?

Plan a Model

Draw or list the materials you will need.

The animals have what they need to make them happy and healthy!

Make a Model

Make your diorama of Ruby's play area.

Draw a picture. Use arrows and labels to show how plants and animals live together.

💬 What did you draw in the play area?
Why? Tell a partner.

Impacts on Earth's Systems

How do you get what you need to live?

FINN
Construction Manager

 ## Science in My World

▶ Watch the video.
You see a living thing building its home.
Think about what it needs to live. What
do you wonder about how it uses things?
Draw a picture.

 ## Science and Engineering Practices

I will engage in argument from evidence.

Plants Change Environments

 **PAGE KEELEY SCIENCE PROBES**

Plants and the Environment

Helen   Jay

Two friends are talking about how living things change the environment. Who has the best idea?

Helen: *Plants can change the environment.*

Jay: *I disagree. Only people can change the environment.*

💬 **Talk about your thinking with a partner.**

Science in My World

You see tree roots breaking a sidewalk. Think about how this happens. What do you wonder?

I wonder how the tree roots broke the sidewalk.

KAYLA
Landscape Architect

? Essential Question
How do plants change the environment?

Cut and paste photos of plants that changed their environment.

Science and Engineering Practices

I will engage in argument from evidence.

 **Name** _____

 Inquiry Activity
Tree vs. Rock

Make a Prediction What will happen to the rock? Draw a picture.

[blank drawing box]

Make a Model

Use clay and twigs to show how tree roots change a rock.

BE CAREFUL. Wash your hands after touching the clay.

Make your tree grow. How do trees get bigger?

Name _____

Record Data What happened to the rock? Draw a picture.

abc Vocabulary

Listen for these words as you learn
about how plants change environments.

environment　　　　　　　need

Plants and Animals Change Their Environments

Read.

Use the new words. Draw a picture.

Changing Environments

▶ Watch the video.
Look at the pictures. Circle the picture
that shows changes made by humans.

Name _____

EXPLAIN

Plants and Animals Change
Their Environments

📖 Read.
Plants make shade. Draw a plant
that makes shade for people.

Copyright © McGraw-Hill Education

Sidewalk Crack

 Explore to learn how the tree changed its environment. Look at the picture. Circle the part the tree changed.

Science and Engineering Practices

Say the **I can** statement together.
Tell a partner what you can do.

I can engage in argument from evidence.

 Name _____

Plants Clean the Air

Explore. Plants give off oxygen.
What living things breathe oxygen?
Circle them below.

Performance Task
Plant Poster

Make a poster about how plants change the environment.

Materials

☐ poster board

☐ markers or colored pencils

Make a Model

1. Write the words **Before** and **After** on your poster board.

Before	After

What kinds of changes do plants make to the environment?

2. Draw a plant under **Before**.

3. Draw an environment.

4. Think about how the plant would change the environment.

5. Draw the changes under **After**.

🗣 Talk about your poster with the class.

? Essential Question

How do plants change the environment?

💬 Talk about the question as a class.
Write the answer.

_____

- -

_____

_____

- -

_____

⚙ Science and Engineering Practices

I did engage in argument from evidence.

Now that you're done with the lesson, rate how well you did.

Rate Yourself

Name _____

Animals Change Environments

PAGE KEELEY
SCIENCE
PROBES

Animals and the Environment

Patti   Dee   Bobby

Three friends are talking about how
living things can change the environment.
Who has the best idea?

Patti: *Only animals can change the environment.*

Dee: *Only plants can change the environment.*

Bobby: *Both plants and animals can change
the environment.*

💬Talk about your thinking with a partner.

 # Science in My World

You see a prairie dog. Think about how its burrow got there. What do you wonder?

I wonder how the prairie dog built its burrow.

CHLOE
Carpenter

? Essential Question
How do animals change the environment?

Draw an animal building its home.

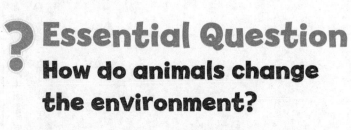

Science and Engineering Practices

I will engage in argument from evidence.

✋ Inquiry Activity
Ant Farm

Make a Prediction What will the ant farm look like next week? Draw a picture.

[blank drawing box]

Carry Out an Investigation

Look carefully at the ant farm every day for one week.

BE CAREFUL. Follow your teacher's instructions. Do not use the hand lens to look at lights or the Sun.

How does the ant farm change?

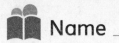
Record Data What does the ant farm look like today? Draw a picture.

Day 1

Day 2

Record Data What does the ant farm look like today? Draw a picture.

Day 3

Day 4

Record Data What does the ant farm look like today? Draw a picture.

Day 5

💬 How did the ant farm change?
   Tell a partner.

Name _____

Moles Change Their Environment

Explore.

Make a Prediction Circle the answer.
Which field will have less grass?

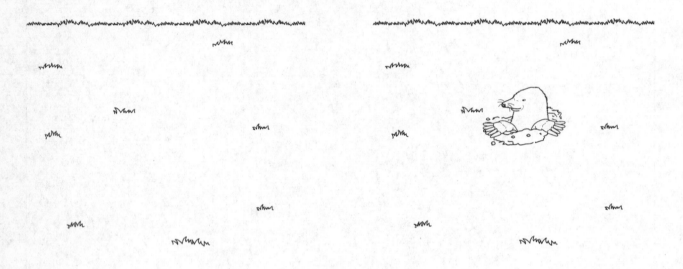

How does the mole find food?
Tell a partner.

Help the mole find grubs.

Record Data Draw the field that has less grass.

Name _____

🔤 Vocabulary

Listen for these words as you learn
about how animals change environments.

burrow   dam

Animals Change Their Environments

📖 Read.

💬 Talk about it. Use the new words.
Draw what you talked about.

Animals Changing Environments

▶ Watch the video. Look at the pictures below. What needs does a beaver have? Circle all the needs a beaver meets when it builds a dam.

a home

sunlight

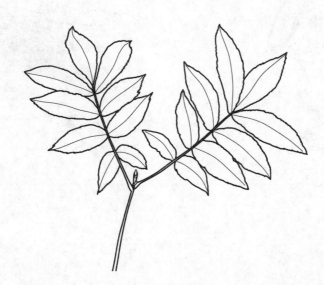

food

Name _____

Animals Change Their Environments

📖 Read.
Draw a line from the animal to
the change it made.

Animals Help Plants

Explore.
Think of some ways animals help plants.
Cut out pictures. Glue them here.

[blank box for glued pictures]

Science and Engineering Practices

How do animals help plants?

Say the **I can** statement together.
Tell a partner what you can do.

I can engage in argument
from evidence.

Name _____

✋ Inquiry Activity
Annoying Animals

Animals change their environments.
Sometimes changes are bad for people.
Think of a bad change an animal makes.
Draw a picture.

💬 Explain your drawing to
the class.

How can people stop animals from making bad changes?

📖 **Name** _____

EVALUATE

⚙️ Performance Task
Beaver Dam

Make a model of a beaver dam.

Carry Out an Investigation

1. Fill the plastic container with sand or mud.

2. Dig a path for the river.

3. Choose a spot to build a dam.

4. Make your dam. Use craft sticks and rocks.

5. Test your dam. Pour water into the river path.

BE CAREFUL. Follow your teacher's instructions. Wash your hands after you have finished.

💬 What happened to the water?

Materials

- [ ] safety goggles
- [ ] plastic container
- [ ] sand or mud
- [ ] craft sticks
- [ ] rocks
- [ ] water

How did the dam affect the river?

Copyright © McGraw-Hill Education

Online Content at connectED.mcgraw-hill.com Lesson 2 Animals Change Environments **157**

Name _____

Record Data What happened to the river?
Draw a picture.

? Essential Question
How do animals change the environment?

🗨 Talk about the question as a class.
Write the answer.

_____

- -

_____

- -

_____

- -

_____

⚙ Science and Engineering Practices

I did engage in argument from evidence.

Now that you're done with the lesson, rate how well you did.

Rate Yourself

Name _____

People Change Environments

People and the Environment

| Haddie | Farrah | Malcolm |

Three friends are talking about how things can change the environment. Who has the best idea?

Haddie: *People change the environment in good ways.*

Farrah: *People change the environment in bad ways.*

Malcolm: *People change the environment in good and bad ways.*

💬 Talk about your thinking with a partner.

Name _____

Science in My World

▶ Watch the video.

You see a family building a tree house. What do you wonder?

I wonder how the family is changing the tree.

KAYLA
Landscape Architect

❓ Essential Question
How do humans change the environment?

💬 Think of a way you have changed your environment. Tell a partner. Draw your idea.

[blank drawing box]

Science and Engineering Practices

I will engage in argument from evidence.

✋ Inquiry Activity
School Changes

Ask a Question

Look at the two maps. How has the area changed?

Record Data Draw the changes.

What kinds of changes did humans make to the environment?

Before	
After	

💬 How did humans change the area?

🔤 Vocabulary

Listen for these words as you learn
about how people change environments.

garden   farm

💬 Talk about it. Use the new words.
Draw a picture.

Name _____

People Changing Environments

▶ Watch the video. Look at the pictures.
Circle the change humans made.

a dam　　　　a farm　　　　a cave

Humans Change Environments

🔊 Explore. Humans plant gardens.
What need does a garden meet?
Circle the answer.

food　　　　shelter　　　　water

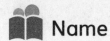
 **Name** _____

Inquiry Activity
Environment Changes

Ask a Question
Draw two ways humans change
the environment. Circle the changes
that were good.

⚙ Crosscutting Concepts
Systems and System Models

💬 How do people get what they need from the environment? Talk to a partner.

⚙ Science and Engineering Practices

Say the **I can** statement together.
Tell a partner what you can do.

I can engage in argument from evidence.

What makes a change good? What makes a change bad?

How Humans Find Food in Their Environment

👁 Look.

Where does your food come from?
Draw a picture.

How does finding food change the environment?

Performance Task
Plan a Garden

Plan a garden with your class.

Make a Model

① Choose a type of garden.

② Choose the location for your garden.

③ Choose the size of your garden.

④ Choose the plants for your garden.

⑤ Help draw the plan for the garden.

💬 Did your class agree? Are you ready to plant the garden? Draw your model.

❓ Essential Question
How do humans change the environment?

🗨 Talk about the question as a class. Write the answer.

_____

- -

_____

- -

_____

- -

_____

Science and Engineering Practices

I did engage in argument from evidence.

Now that you're done with the lesson, rate how well you did.

Rate Yourself

Impacts on Earth's Systems

⚙ Performance Project
Create a Poster

Use what you have learned to show how plants and animals can change their environment.

1 Divide your poster in two.

2 Label one side **Before**. Label the other side **After**.

3 Under **Before**, draw one environment.

4 Under **After**, draw how the environment changed.

5 Draw the needs the plant or animal met.

Plants, animals, and people change their environments to get what they need to live.

Explain how the environment was changed.
Draw the needs the plant or animal met.

💬 What needs are met by a plant or animal's habitat? What happens when these needs are not met by the habitat?

Protecting Our Earth

How can I keep people from polluting the park?

POPPY
Park Ranger

🌍 Science in My World

You see the bottles. What do you wonder about how the bottles were used?
Draw a picture.

⚙️ Science and Engineering Practices

I will obtain information.
I will evaluate information.
I will communicate information.

Land, Air, and Water Pollution

 **PAGE KEELEY SCIENCE PROBES** **Pollution**

Joanie   Liam

Two friends are talking about how people pollute the land, air, and water. Who do you think has the best idea about what people can do about polluting the environment?

Joanie: *I think people can stop pollution.*

Liam: *I think people can reduce pollution.*

🗪 **Talk about your thinking with a partner.**

Name _____

 # Science in My World

You see trash. Think about how it got there. What do you wonder about the living things in the water?

I wonder if the pollution will hurt the fish.

KAYLA
Landscape Architect

? Essential Question

How do people's actions change land, air, and water?

Land	Air	Water

Science and Engineering Practices

I will obtain information.
I will evaluate information.

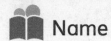
 Inquiry Activity
River Pollution

How does the pollution change the water?

Make a Prediction What can be affected by water pollution? Circle the answer.

A.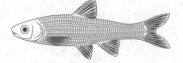

B.

C.

D.

E.

F.

Carry Out an Investigation

Add pollutants to the water. Watch carefully.

Name _____

Record Data Draw how the water looked.

What would happen to plants or animals that tried to live in this water?

💬 Tell a partner how the water looked.

🔤 Vocabulary

Listen for this word as you learn
about land, air, and water pollution.

pollution

A Big Difference

📖 Read.

🗨 Talk about it. Use the new word.

Pollution in the Park

🔊 Explore the park. Draw the kinds of
pollution you found in the park.

🗨 How can you reduce pollution in
the park? Tell a partner.

Humans Affect Earth

Read.

What is waste? Tell a partner.

Oil Spill

Watch the video.

What animals will be hurt by an oil spill?

Crosscutting Concepts
Cause and Effect

Draw the effect of an oil spill.

Science and Engineering Practices

I can obtain information.
I can evaluate information.

Inquiry Activity
Listen to the Story

Draw Conclusions What will happen after the boy plants the seed? Draw a picture.

[ ]

💬 Talk with a partner about the story. What types of pollution were in the story?

Inquiry Activity
Oil and Water

Mix the oil and water a little, then a lot.

Investigate how oil acts in water.

Carry Out an Investigation

Add water and oil to a bag.
Close the bag. Mix them together.

Record Data Draw what you see.
Color the water blue and the oil yellow.

Before mixing	After mixing

What did you see when you mixed the oil and water? Tell a partner.

 Name _____

Performance Task
Air Quality

Materials

- [ ] petroleum jelly
- [ ] index cards
- [ ] plastic spoon

Directions

1 Spread jelly on the cards.

2 Leave one card inside.

3 Leave one card outside.

4 Check the cards after one day.

5 Check them after two days.

6 Draw what you see on the next page.

Watch the cards for two days.

Name _____

Carry Out an Investigation

Record Data Draw what you see.

Day	Card 1	Card 2
After 1 day		
After 2 days		

💬 Talk with a partner. How can land, air, and water become polluted?

? Essential Question

How do people's actions change land, air, and water?

💬 Talk about the question as a class.
Write the answer.

_____

_____

_____

_____

_____

_____

⚙ Science and Engineering Practices

I did obtain information.
I did evaluate information.

Now that you're done with the lesson, rate how well you did.

Rate Yourself

Help Save Natural Resources

PAGE KEELEY SCIENCE PROBES

Natural Resources

Felix

Larry

Two friends have different ideas about natural resources. Who do you agree with?

Felix: *I think natural resources are things made out of natural materials.*

Larry: *I think natural resources are things in nature that people use.*

💬 Talk about your thinking with a partner.

 Science in My World

▶ Watch the video. See the water from the tap. What do you wonder about the water?

> I wonder how we protect natural resources like water.

JIN
Paleontologist

? Essential Question
How can we save natural resources?

Draw two natural resources.

Science and Engineering Practices

I will obtain information.
I will evaluate information.

✋ Inquiry Activity
Use Less Water

Make a Prediction How can we use less water? (Circle) one.

Should you let the water run while you brush your teeth and wash the dishes?

Carry Out an Investigation

Plug the sink. Wash the dishes. How much water do you use?

Record Data Draw what you saw each time you washed the dishes.

First Washing	Second Washing

💬 Should you leave the water running or turn it off? Why? Tell a partner.

Name _____

abc Vocabulary

Listen for these words as you learn
about how to save natural resources.

conserve   natural resource

Natural Resources

📖 Read.

💬 Talk about it. Use the new words.
Draw a picture.

Inquiry Activity
Natural Resources in the Classroom

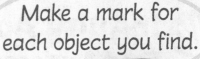
Make a mark for each object you find.

Carry Out an Investigation

Record Data Count things made from plants and from animals. Use tally marks.

Plants	Animals

💬 Talk to a partner. Were more things made from plants or animals?

⚙️ Science and Engineering Practices

I can obtain information.
I can evaluate information.

Name _____

✋ Inquiry Activity
Food Sources

Make a Prediction Where do you think your food comes from? (Circle) ways people get food.

Name _____

Carry Out an Investigation

Draw your favorite food. Where do you think your food comes from?

💬 Tell a partner about your favorite food.

Favorite Food	Natural Resource

Where does each food come from?

Performance Task
Help the Environment

Materials

☐ paper

☐ drawing pencils

Make a Model

1 Think of a natural resource.

2 Think of a way to use less of the resource.

3 Make a model of your idea.

💬 Tell a partner your ideas.

? Essential Question
How can we save natural resources?

💬 Talk about the question as a class.
Write the answer.

_____
- -
_____
- -
_____
- -
_____

⚙️ Science and Engineering Practices

I **did** obtain information.
I **did** evaluate information.

Now that you're done with the lesson, rate how well you did.

Rate Yourself

Reduce, Reuse, Recycle

 PAGE KEELEY SCIENCE PROBES

Reduce, Reuse, Recycle

Marcus

Mandy

Two friends are looking at a 3Rs poster. They each have a different idea about the 3Rs. Who do you agree with?

Mandy: *I think reduce, reuse, and recycle is about not having any trash.*

Marcus: *I think reduce, reuse, and recycle is about having less trash.*

💬 **Talk about your thinking with a partner.**

Name _____

Science in My World

▶ Watch the video. What do you wonder about the objects?

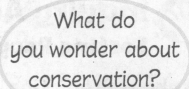

What do you wonder about conservation?

❓ Essential Question
How can we take care of Earth?

What are some things you can do to take care of Earth? Draw a picture.

KAYLA
Landscape Architect

Science and Engineering Practices

I will obtain information.
I will communicate information.

Inquiry Activity
Sorting Recyclables

Make a Prediction What items can be recycled? (Circle) your ideas.

Label five containers: trash, plastic, paper, metal, and other. Sort the trash.

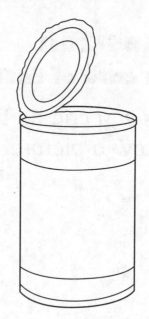

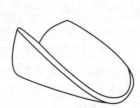

Carry Out an Investigation

Collect the trash from your classroom for one week.

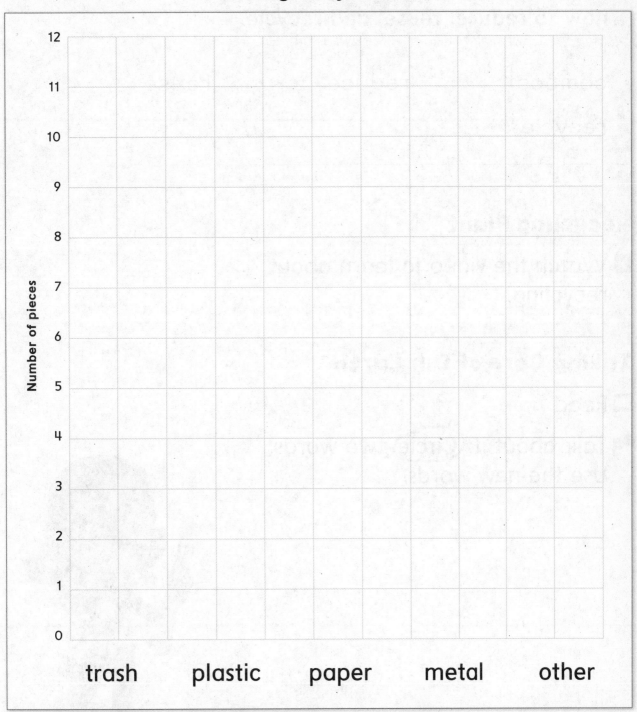

Name _____

Record Data Collect trash for one week.
Sort it. Make a graph.

Sorting Recyclables

Number of pieces

trash   plastic   paper   metal   other

💬 Talk about it. Did you have more trash,
plastic, paper, or metal?

Name _____

🔤 Vocabulary

Listen for these words as you learn
how to reduce, reuse, and recycle.

compost   reduce   reuse

recycle

Recycling Plant

▶ Watch the video to learn about
recycling.

Taking Care of Our Earth

📖 Read.

💬 Talk about it. (Circle) two words.
Use the new words.

Draw things you can reduce, reuse, or recycle.

Reduce	
Reuse	
Recycle	

Recycled Objects

📷 Explore. How do you make new things from old things?

⚙ Crosscutting Concepts
Cause and Effect

How does recycling change a material?

Material	After Recycling

⚙ Science and Engineering Practices

Say the **I can** statement together.
Tell a partner what you can do.

I can obtain information.
I can communicate information.

Composting

🔊 Explore to learn about composting.

Make a composter. Draw a picture of the container. Draw what you would put inside.

Name _____

⚙ Performance Task
Reduce Trash Poster

Work with a group. Think of three ways to make less trash in your classroom. Make a poster to share your ideas.

<table>
<tr><td colspan="2">Materials</td></tr>
<tr><td>☐</td><td>paper</td></tr>
<tr><td>☐</td><td>poster board</td></tr>
<tr><td>☐</td><td>markers</td></tr>
<tr><td>☐</td><td>glue</td></tr>
<tr><td>☐</td><td>scissors</td></tr>
</table>

Make a Model

1 Draw three ways to reduce trash.

2 Color your pictures.

3 Label your poster.

💬 Talk to a partner. What ways could you make less trash each week?

	Reduce Trash
1	
2	
3	

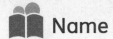
? Essential Question

How can we take care of Earth?

🗩 Talk about the question as a class.
Write the answer below.

_____
- -
_____
- -
_____
- -
_____
_____

⚙ Science and Engineering Practices

I did obtain information.
I did communicate information.

Now that you're done with the lesson, rate how well you did.

Rate Yourself

Name _____

Protecting Our Earth

⚙ Performance Project
What Natural Resources Do You Use?

What natural resources does your family use? Make a plan to reduce how much your family uses.

Plan an Investigation

Draw or list the materials you will need.

What natural resources do you use?

Design a Solution

What can you do to reduce your use of a natural resource? Draw your plan.

💬 **What happens if a natural resource is polluted?**

An Interview with
Dinah Zike Explaining
Visual Kinesthetic Vocabulary®, or VKVs®

What are VKVs and who needs them?

" VKVs are flashcards that animate words by kinesthetically focusing on their structure, use, and meaning. VKVs are beneficial not only to students learning the specialized vocabulary of a content area, but also to students learning the vocabulary of a second language. "

Dinah Zike | Educational Consultant
Dinah-Might Activities, Inc. – San Antonio, Texas

Why did you invent VKVs?

" Twenty years ago, I began designing flashcards that would accomplish the same thing with academic vocabulary and cognates that Foldables® do with general information, concepts, and ideas—make them a visual, kinesthetic, and memorable experience. "

I had three goals in mind:

- **Making two-dimensional flashcards three-dimensional**

- **Designing flashcards that allow one or more parts of a word or phrase to be manipulated and changed to form numerous terms based upon a commonality**

- **Using one sheet or strip of paper to make purposefully shaped flashcards that were neither glued nor stapled, but could be folded to the same height, making them easy to stack and store**

Dinah Zike's
VKV
Visual
Kinesthetic
Vocabulary®

Why are VKVs important in today's classroom?

" At the beginning of this century, research and reports indicated the importance of vocabulary to overall academic achievement. This research resulted in a more comprehensive teaching of academic vocabulary and a focus on the use of cognates to help students learn a second language. Teachers know the importance of using a variety of strategies to teach vocabulary to a diverse population of students. VKVs function as one of those strategies.

An Interview with
Dinah Zike Explaining
Visual Kinesthetic Vocabulary®, or VKVs®

Dinah Zike's
Visual
Kinesthetic
Vocabulary®

How are VKVs used to teach content vocabulary?

" As an example, let's look at content terms based upon the combining form *–vore*. Within a unit of study, students might use a VKV to kinesthetically and visually interact with the terms *herbivore*, *carnivore*, and *omnivore*. Students note that *–vore* is common to all three words and it means "one that eats" meat, plants, or both depending on the root word that precedes it on the VKV. When the term *insectivore* is introduced in a classroom discussion, students have a foundation for understanding the term based upon their VKV experiences. And hopefully, if students encounter the term *frugivore* at some point in their future, they will still relate the *–vore* to diet, and possibly use the context of the word's use to determine it relates to a diet of fruit. "

What organization and usage hints would you give teachers using VKVs?

" Cut off the flap of a 6" x 9" envelope and slightly widen the envelope's opening by cutting away a shallow V or half circle on one side only. Glue the non-cut side of the envelope into the front or back of student workbooks or journals. VKVs can be stored in the pocket.

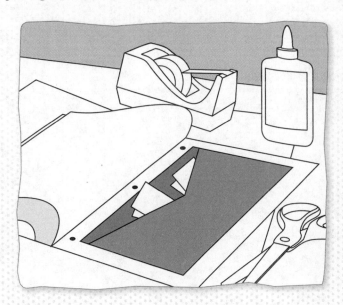

Encourage students to individualize their flashcards by writing notes, sketching diagrams, recording examples, forming plurals (radius: radii or radiuses), and noting when the math terms presented are homophones (sine/sign) or contain root words or combining forms (kilo-, milli-, tri-).

As students make and use the flashcards included in this text, they will learn how to design their own VKVs. Provide time for students to design, create, and share their flashcards with classmates. "

Dinah Zike's book Foldables, Notebook Foldables, & VKVs for Spelling and Vocabulary 4th-12th won a Teachers' Choice Award in 2011 for "instructional value, ease of use, quality, and innovation"; it has become a popular methods resource for teaching and learning vocabulary.

distance

Distance is the amount of space between two places or things.

A **pull** is a force that moves something closer to you.

The students gave the rope a pull.

A **push** is a force that moves something away from you.

The girls push the door open.

✂ cut on all dashed lines   ⬜ fold on all solid lines

t

pull

push

Memory Maker: Draw a story using the words on this VKV.

Memory Maker: Draw yourself at a large distance from school.

VKV210 Forces and Motion

✂ cut on all dashed lines   fold on all solid lines

The **planet** we live on is called Earth.

When an object blocks light from the **Sun** it makes **shade.**

Shade is the dark area caused when light is blocked.

The **Earth** is the third planet from the Sun.

Earth

The **Sun** is the star closest to Earth.

Dinah Zike's
Visual
Kinesthetic
Vocabulary®

Energy and the Sun

✂ cut on all dashed lines   ▭ fold on all solid lines

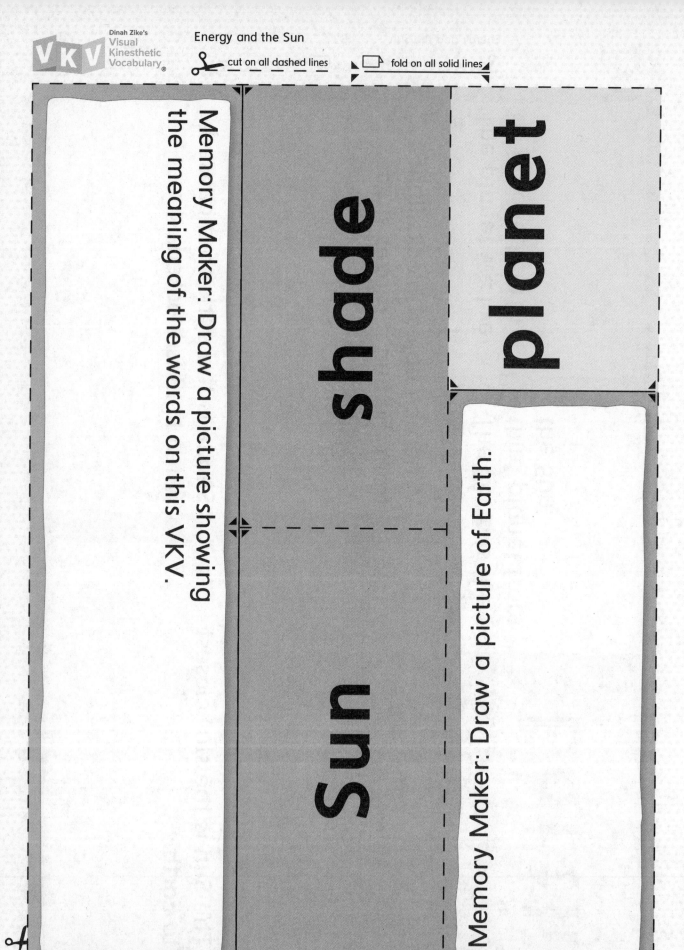

Memory Maker: Draw a picture showing the meaning of the words on this VKV.

shade

planet

Sun

Memory Maker: Draw a picture of Earth.

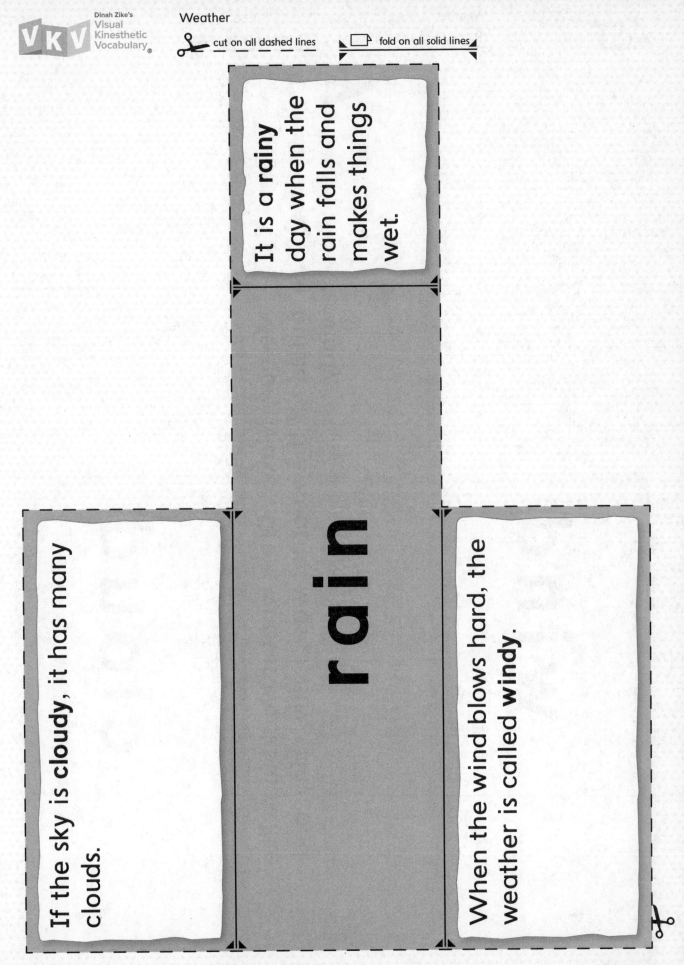

cut on all dashed lines

fold on all solid lines

It is a rainy day when the rain falls and makes things wet.

rain

If the sky is cloudy, it has many clouds.

When the wind blows hard, the weather is called windy.

Dinah Zike's
Visual
Kinesthetic
Vocabulary®

✂ cut on all dashed lines

◻ fold on all solid lines

y

Memory Maker: Draw what it would look like outside if the weather was cloudy, rainy, and windy.

wind

cloud

Dinah Zike's
Visual
Kinesthetic
Vocabulary®

✂ cut on all dashed lines

📄 fold on all solid lines

fall season

1. Summer is the time of year after spring.

2. It is hottest in summer.

1. Spring is the season after winter.

2. Fall is the season after summer.

1. Winter is the time of year after fall.

2. It is coldest in winter.

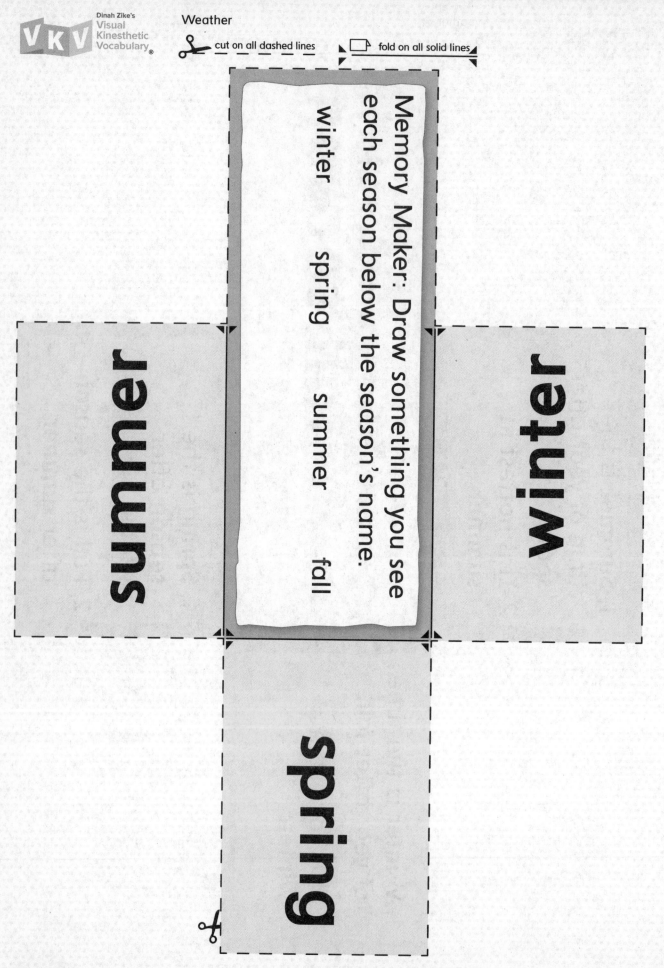

Dinah Zike's
Visual
Kinesthetic
Vocabulary®

cut on all dashed lines

fold on all solid lines

Memory Maker: Draw something you see each season below the season's name.

winter   spring   summer   fall

summer

winter

spring

Weather

✂ cut on all dashed lines

▭ fold on all solid lines

forecast

The **weather** is what the sky and air are like each day.

Memory Maker: Draw a weather forecast showing weather patterns.

✂ cut on all dashed lines   fold on all solid lines

weather patterns

1. **Patterns** are the repeated way in which something happens.

2. Weather **patterns** are the repeated way the **weather** happens.

To **forecast** means to say that something will happen by using information about the **weather**.

VKV **Dinah Zike's**
Visual
Kinesthetic
Vocabulary®

Weather

✂ cut on all dashed lines

▭ fold on all solid lines

1. A **sunny** day is full of sunlight.

2. Underline the word you see in both *sunny* and *sunlight.*

The **Sun** is the star closest to **Earth.**

Severe weather means there are very strong conditions outside.

severe weather

Sun

✂ cut on all dashed lines

▱ fold on all solid lines

sunny

Memory Maker: Draw what severe weather looks like.

The **weather** is what the sky and air are like each day.

Memory Maker: Draw a sunny day.

VKV220 Weather

VKV Dinah Zike's
Visual
Kinesthetic
Vocabulary®

Weather

✂ cut on all dashed lines

fold on all solid lines

A **thunderstorm** is a storm with thunder, lightning, rain, and wind.

A **hurricane** is a strong storm with winds that form a cloud that looks like a funnel.

storm

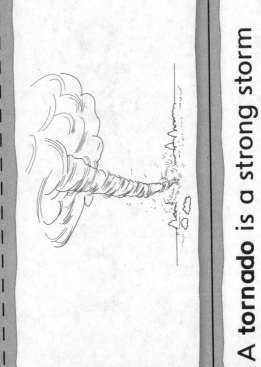

A **tornado** is a strong storm with heavy rain and winds that blow in a circle.

Dinah Zike's
VKV
Visual
Kinesthetic
Vocabulary®

Weather

✂ cut on all dashed lines | ▱ fold on all solid lines

Memory Maker: How is a tornado different from a hurricane? Draw or write your answer.

hurricane

thunder

tornado

Memory Maker: Draw what happens during a thunderstorm.

A nonliving thing is a thing that does not grow and change, or need food, air, or water to survive.

A living thing is something that grows, changes and needs food, air, and water to survive.

living

Draw an animal.

Draw a plant.

An animal is a living thing that is not a human or a plant.

A plant is a living thing that has leaves, roots, and makes its own food.

Dinah Zike's
Visual
Kinesthetic
Vocabulary®

✂ cut on all dashed lines

📄 fold on all solid lines

non-

animal

plant

Memory Maker: Draw a circle around the animal and a square around the plant.

Memory Maker: Draw pictures of a nonliving and a living thing.

cycle

Air is the invisible gas that people and animals breathe.

Soil is the top layer of Earth.

A **life cycle** is how a living thing grows, lives, and dies.

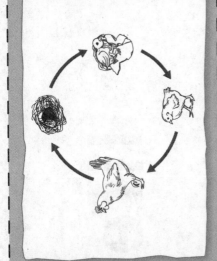

Water is the liquid that falls as rain from the sky.

Dinah Zike's
Visual Kinesthetic Vocabulary®

✂ cut on all dashed lines

▭ fold on all solid lines

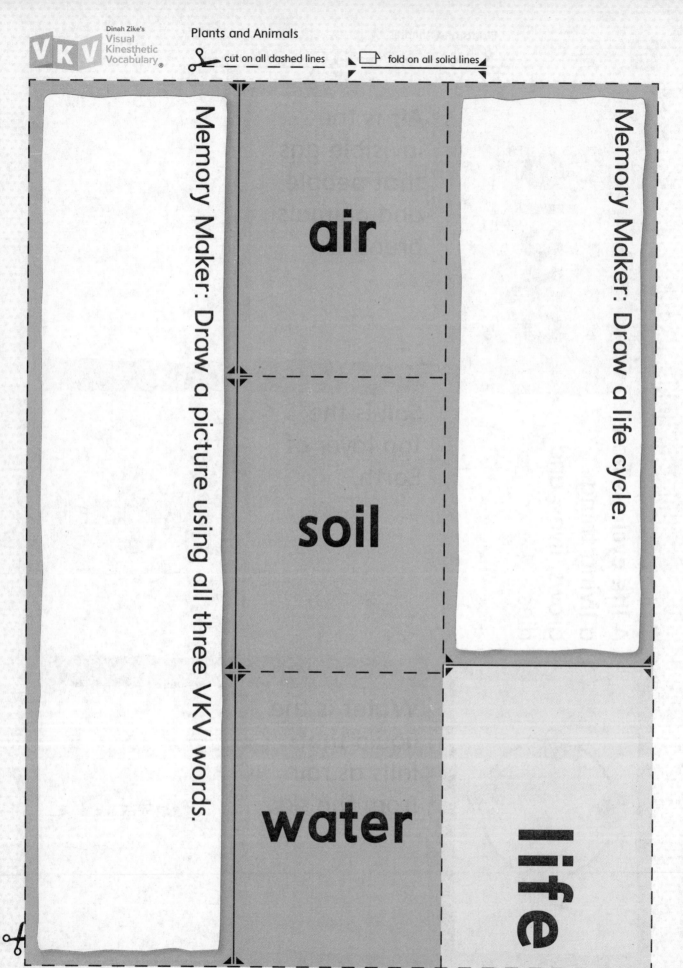

Memory Maker: Draw a picture using all three VKV words.

Memory Maker: Draw a life cycle.

air

soil

water

life

Dinah Zike's
Visual
Kinesthetic
Vocabulary®

✂ cut on all dashed lines

▭▸ fold on all solid lines

Circle the animals that live in burrows.

A burrow is a hole or tunnel in the ground that an animal makes to live in.

A garden is an area of land used for growing flowers or vegetables.

burrow

flower garden

Dinah Zike's
Visual
Kinesthetic
Vocabulary®

✂ cut on all dashed lines

🗀 fold on all solid lines

ing

Memory Maker: Draw the burrow you want to live in if you are a burrowing animal.

Memory Maker: Create your own garden, include as many flowers or vegetables as you like.

vegetable

Dinah Zike's
Visual
Kinesthetic
Vocabulary®

VKV

✂ cut on all dashed lines

⬜ fold on all solid lines

Pollution is anything harmful in the air, land, or water.

Compost is a mixture of dead plants.

To **conserve** means to save, keep, or protect something.

pollution

compost

conserve

Dinah Zike's
Visual
Kinesthetic
Vocabulary ®

VKV

ation

e

e

de

Memory Maker: Draw a picture of a pile of compost for a garden.

Memory Maker: Draw a picture that will make people want to conserve trees.

Memory Maker: Draw pollution you have seen in your town.

Dinah Zike's
**Visual
Kinesthetic
Vocabulary**®

✂ cut on all dashed lines

📠 fold on all solid lines

recycle

1. **Reuse** means to use something again.
2. **Recycle** means to make something new from something old.

Reduce means to use less of something.

Dinah Zike's
Visual
Kinesthetic
Vocabulary®

Protecting Our Earth

✂ cut on all dashed lines

fold on all solid lines

duce

use

Memory Maker: Draw a comic strip showing the words on this VKV.